Productivity

Become a Master in Getting Things Done

Copyright © 2017 A.C. Drexel

All rights reserved.

ISBN: 1974277682
ISBN-13: 978-1974277681

© Copyright 2017 - All rights reserved.
The contents of this book may not be reproduced, duplicated or transmitted without direct written permission from the author.

Under no circumstances will any legal responsibility or blame be held against the publisher for any reparation, damages, or monetary loss due to the information herein, either directly or indirectly.

Legal Notice:
This book is copyright protected. This is only for personal use. You cannot amend, distribute, sell, use, quote or paraphrase any part or the content within this book without the consent of the author.

Disclaimer Notice:
Please note the information contained within this document is for educational and entertainment purposes only. Every attempt has been made to provide accurate, up to date and reliable complete information. No warranties of any kind are expressed or implied. Readers acknowledge that the author is not engaging in the rendering of legal, financial, medical or professional advice. The content of this book has been derived from various sources. Please consult a licensed professional before attempting any techniques outlined in this book.

By reading this document, the reader agrees that under no circumstances are is the author responsible for any losses, direct or indirect, which are incurred as a result of the use of information contained within this document, including, but not limited to, —errors, omissions, or inaccuracies.

ALSO BY A.C. DREXEL

Emotional Intelligence
How to Master your Emotions, Build Self-Confidence and Program Yourself for Success

Free
A Guide to Understand and Set Yourself Free from Co-dependency

Habits
Easy Habits for a Better Life

How to Analyze People
The Ultimate Guide to Human Psychology, Body Language, Personality Types and Ultimately Reading People

Anger Management
How to Control Anger, Develop Self-Control and Ultimately Master Your Emotions

Self Discipline
How to Achieve Mental Toughness, Motivate Yourself and Develop Self Discipline for Life

CONTENTS

	INTRODUCTION	i
1	YOUR GOALS IN LIFE	1
2	LEARNING TO USE YOUR TIME WISELY	4
3	PERSONAL LIFESTYLE HABITS	8
4	HABITS TO AVOID IN ORDER TO BECOME MORE PRODUCTIVE	11
5	GIVING YOURSELF MOTIVATION	14
	CONCLUSION	ii

INTRODUCTION

Have you ever heard the expression "The road to hell is paved by good intentions" and wondered what that means in relationship to your life? The fact is that the number of people suffering from stress in this day and age actually correlates with the amount of people that can't get what they want out of life. If you have looked beyond the cover of this book, you are obviously looking for something. Are you a writer who never actually writes? Are you a salesman who doesn't quite get to his quota every month? Do you have intentions that you never fulfill?

There are several reasons for this and these are covered in this book. What gives me the right to dictate how you can become more productive? I learned through trial and error but I also used very great examples as those whose advice was worth listening to. If you want success, then who better to examine than successful people? The fact is that you may even be sabotaging your own chances of success for various psychological reasons. This book will explain them to you and you can use it as your 'go to' guide to help you to overcome the obstacles that stop you from becoming productive.

I now achieve ten times more than I used to. I have a successful business and I am able to have my fair share of leisure. Why? Because when you manage productivity, you also manage to get your life in the kind of order that allows you to have fun too. People who are successful may initially slave over the process that took them there, but their continued success isn't based upon this idea of keeping their nose to the grind. It's all about using time in a productive way.

There are so many people out there with so much potential they will never reach. The shame is that many of these could be successful in their own chosen field, but what holds them back from success is their own attitude and their psychological blocks. We will be talking about these in the following chapters and helping you to unlock your true potential. It is a potential that is there for everyone in all fields of work. Thus, I hope that what I have written will spur you on to change your attitudes and become as productive in your life as possible. The problem is that life passes all too quickly. You need to do something and do it now. Stop thinking about it. Stop planning it and start actually doing it. This book enables you to do that.

CHAPTER 1 – YOUR GOALS IN LIFE

You need to have clear ideas of what your goals are. This could mean goals for the day, goals for the week, the month or the year. When I asked someone who was not productive about what he intended to achieve this week, he was so stressed out that he could not give me an answer. He had allowed work to pile up to such an extent that the task became enormous.

Why it's necessary to have goals, even small ones, is so that you achieve and can feel good about your achievements. Of course, things may get in the way occasionally but that's normal too. These are called hurdles and everyone experiences them. The way that you minimize hurdles is what matters or the way that you use mistakes to learn from and then move on in a positive way. The particular man that I talked to was always bogged down with work. He didn't eat a nutritious meal at lunchtime. He didn't have much of a family life. His life revolved around his own chaos and the problem with living like this is that you take all the motivation out of life. If you can't measure what you have achieved, how do you expect to achieve more than that on another day? The measurements that you use are very simple. You create a short list of things that you want to achieve and then achieve them.

It sounds very simple when it's said like that, but it really can work like that. Don't be a perfectionist and spend all of your time on detailed listing. Don't be afraid of your own failure and not even try. Both of these failing will stop you from every achieving and people don't know it, but often their own psychological workings are what stands in the way of success. Why make things more complex than they need to be? Why be afraid of failure when you haven't even started your tasks yet?

PRODUCTIVITY

Write down in a notebook what you want to achieve in your life. Close your eyes for a moment and be the success that you always wanted to be. To be successful and productive you have to think like people who are successful and productive. Thus, you need to set your mindset to success, rather than sit on your laurels and believe that success happens to everyone else except you.

I once asked a friend of mine what he wanted in life and he really didn't have much direction at all. Be ambitious, picture what you want, but unless you are prepared to do so, your life will be mediocre. The reason for this is that it's a little like setting out in a car and not really knowing where you are going. Your notes act as your map in life and when you have ideas about what you want to be in life, you are much more likely to achieve your goals.

The next task in this chapter is to write down your plans for the week and your plans for the next day. When you make goals that are not attainable or try to push yourself further than hours permit, what you end up with is disappointment. Thus, you need to be able to triage your work and decide upon the following:

- What work can wait or is low priority?
- What work needs your full attention and is relatively urgent?
- What work can you give to someone else?

When you know what work needs to be done within the next 24 hours, you may look at your list and despair because it's all too much for you. However, there's a very good way to deal with it once you have sorted out what you can achieve in the next 24 hours. The work that is routine and non-urgent can be put to one side for the morning tomorrow. The work that you can delegate to someone else shows trust in another member of staff. Stop being the martyr when you know that others can do certain tasks better than you can. Let them. If you empower other people, you show your ability to manage and that's why some people get promoted quicker than others. It isn't about YOU doing all the work yourself. No one likes martyrs. However, if you delegate and show yourself to be a team leader, then you are much more likely to be considered as potential management material.

Now create the list of the urgent work and if this still seems hard, split large jobs into smaller and more manageable portions, so that each of these parts of the job can be considered as goals. What you are doing is making your goals more manageable and that's what helps you when you are trying to be productive. When you see success, albeit small in the first couple of days, this gives you the motivation to do better and you will find that you will actually achieve much more than you ever did before. Challenging yourself is a good thing, though in the early days of giving yourself lists, make the lists doable so that you have the incentive to succeed in the future and then make the lists a little harder so that you have to work to achieve them. This increases your level of satisfaction and also your production rate.

CHAPTER 2 – LEARNING TO USE YOUR TIME WISELY

How many times are you interrupted while trying to do a job? How many times does the phone go? How many times do you stop to look at emails? How many inquiries do you get from other members of staff? The fact is that a huge percentage of working time is lost because of multi-tasking. When you are in the middle of a complex problem and allow yourself to be interrupted, what happens is that your mind takes a while to get back to the initial problem and all that time is wasted.

Silent time

The most productive time for your brain is first thing in the morning and just after lunch. Your mind is rested and you are able to think straight. Thus, you need to create a silent time. I found that working outside normal working hours, I achieved ten times as much work than I did while the office doors were open to the public. While I am not suggesting that you give away your free time as I did, what I am suggesting is that you put aside a couple of hours each morning for the tasks that are difficult and that require your full attention. During this time, you need to:

- Switch off your cell phone.
- Put your office phone onto voicemail
- Put a 'Do not disturb' sign on your door
- Put your emails into auto response mode
- Switch off the internet.

If you think that people in the workplace will not respect your wishes, talk about this silent time with your boss and explain that you need this time to get up to speed and to give your clients your full attention. No boss worth his salt will deny you this because he will know that everyone needs downtime to get on with the jobs that they have to do.

During this time, you deal with the difficult jobs of the day and the more difficult the job, the sooner you should do it. When you do the most difficult tasks first, what you are doing is making the rest of your day easier. Use your mind when it is in optimal mode first thing in the morning or after lunch and let it get through the work that it needs to do. The kind of work that can wait until later is the more routine work. Answering emails is pretty easy to do even if you have people around you, but you need to learn the rules when it comes to multi-tasking. If you are in the middle of writing an email and someone comes to your desk, signal that person to wait for a moment and finish the email. That way, you don't lose your train of thought and you finish each task that you start.

Working out your priorities

The Pareto Principle

You may not know it but there is a force that works against us all of the time and that is called the Pareto Principle. A long time ago in Italy, an architect by the name of Pareto worked out that 80 percent of the population worked hard for the benefit of 20 percent. That doesn't seem right does it? However, this principle rings true in all areas of your life including your work, your home life and everything that you do. What you need to learn to do is to swing the balance in your favor. Thus, you need to work out how much time you spend with your family and friends and how much is spent working. Do you split the two areas or do you take your work home? You need to draw a pair of scales onto a piece of paper and every time that you do a task during the day you choose to examine your habits, write a habit onto a sticky note and place it on the family side or the working side of the scales. I would not mind betting that your work life gets the 80 percent of your time and that you have put your family and friends on a backburner in an effort to move forward with your life.

The problem with this is that you eventually end up doing more that you don't enjoy than that which you do enjoy and you need to swing the balance a bit in your favor. For example, if you come home exhausted every night of the week, you are not going to do your best in your job the next day and your family are not going to get the best out of you. Thus, make sure that you have sufficient down time to enjoy yourself so that you build your energy and can enjoy your work life more. If you are happy and content with the balance, you work harder and thus achieve more, but you also please your family and friends and enjoy your time off more. Look at the areas of your life which are causing you the most pain in life and try to adjust the balance so that you get more good things in your life. Once you do, you will be able to produce more and be happy to do that.

Delegation

Good managers of people know how to delegate and they also know the significance of doing this. It helps members of staff to work together toward the same aim. It also means putting trust in someone else rather than thinking you are the only person who can do a particular job. Stop job-hogging because it doesn't make you look more efficient. In fact, it alienates other people. Trust them and when you receive emails that you know someone else is more capable of answering quicker than you, forward the email with a c.c. that is sent back to you that can be placed in a diary box in your email account, so that you can check progress at a later date.

When you delegate work to someone else, that person needs to know what the deadline is because how else will you know that the job has been done. You can also ask for progress reports and tell that person that if there are any delays or, if he has any difficulty, to keep you advised.

By working out all of your priorities and dividing large jobs into manageable smaller ones, you are able to up your productive time. If you then use special no interruption times to concentrate on harder tasks, you are able to achieve more and get all of the really difficult work out of your in tray by lunchtime. That's quite an achievement.

As you go through the tasks that you mapped out on your list for the day, mark them with a huge red tick once they are fully achieved because as you see the list go down and your desk become emptier, this gives you the incentive to carry on with the whole of your list so that at the end of the day, there is nothing left to do. Pat yourself on the back when you achieve that because it's a wonderful feeling, but don't forget people in your office who don't know how to prioritize. Help out whenever you can as this is what people with management mentality do.

The other thing that you can do to keep yourself motivated is give yourself some kind of prize for finishing a set task. I had a whole heap of work to do and didn't think I was capable of finishing it until I decided to delegate certain portions of it and concentrate on the rest myself. I set a deadline for all people working on that particular job and we kind of competed and this competition makes you even more conscious of the passing of time. If you are delegating and you are all working toward the same deadline, there can be some kind of reward between you when the job is finished and this makes people you work with even more likely to want to help you in the future. You are not putting pressure on them. You are encouraging them to put pressure on themselves to beat you at whatever it is that you have all set out to do.

CHAPTER 3 – PERSONAL LIFESTYLE HABITS

There are some very important lifestyle habits that help you to be productive and you should take these seriously. For example, do you go to bed late every night and expect your mind to work overtime in the morning? What about the food you eat? Do you eat a balanced diet? How do you expect your body to perform at optimal levels if you don't give it the right kind of fuel?

What about relaxation time? Do you give yourself sufficient down time? Everyone needs this and it's important that you don't become 'all work and no play' because if you do, you start to lose your purpose in life and work seems like a drag. You need to make sure that you get sufficient exercise as well because this keeps you on your toes.

Taking time to breathe correctly

There is a very good reason why people take up meditation. It enables them to balance the influences that come into their minds and learn to breathe in a certain way that encourages the right amount of oxygen to be circulating in the bloodstream. If you work on overload all of the time, maybe devoting 20 minutes of your day to meditation would be the most productive thing that you can do because the clarity that you get from meditation makes it a worthwhile exercise.

If you are worried about packing enough hours into the day, set your alarm an hour earlier and do your meditation before breakfast. This helps you to greet the day with a balanced mind and a calm approach. You may not be aware of it but most people only use the top part of the lungs when they breathe. This means that oxygen flow is affected and it also means that panic can set in. Meditation puts this right by teaching you a method of deep breathing that encourages your sympathetic nervous system to carry out all of its functions in your body. A more efficient body means a more efficient mind.

To meditate, you simply sit in a hard chair and keep your back straight. Your feet should be flat on the floor. Your hands are placed one on top of the other – palms upward and thumbs touching. Breathe in through the nostrils to the count of 8, hold the breath to the count of 4 and breathe out to the count of 10. You repeat this process over and over again and at the same time, you try to keep thoughts from entering your head. When they do, you simply observe them as if they were not part of your life and then let them go. Do not attach any emotions to these thoughts because they are not given credence until you do. Learning to let go is one of the best disciplines that you can encompass into your life because it serves you very well in the workplace too. It also makes you a better listener and you may pick up some very valuable ideas along the way.

When you meditate, you allow your subconscious mind to rest and this makes it sharper when it needs to be, later on in the workplace. Meditation has many other benefits such as lowering the blood pressure, reducing the speed of the heartbeat and allowing you to approach the day in a positive frame of mind. The practice needs to be incorporated into your life as a matter of course, each and every day because the more you do it, the better you become at doing it and it really will change your views in life to such an extent that you will feel able to conquer all of the workload that you have.

Drinking water

I am never sure why so many people leave this very important ingredient out of their lives. You need to drink about 8 glasses of water a day and if you start the day right and begin early, you will find that you keep your body hydrated and that you tend to feel healthier when you do this. Many people who suffer from stress do so because the body is trying to tell them that it needs water. This isn't just when you are hot. It isn't just when you are thirsty. Keeping the body able to move and feel good without the onset of inflammation also helps to keep your brain sharp and ready to take on the tasks of the day. If you are a coffee drinker, then you probably know that coffee acts as a diuretic and makes you go to the loo more than water does. This means that for all the liquid intake that you have, you are losing rather a lot. Drinking water will help to hydrate your body and this in turn helps you to be able to feel energetic and be able to do more.

Sleep

If you spend all of your waking hours worried about your workload, then you won't achieve very much. It's far better to make sure that you get your eight hours of sleep. The release of hormones into the body while you sleep helps to mend whatever is not working correctly in your body. It refreshes the mind and helps your concentration levels. Thus, sleep is vital. Put away your computer at a reasonable hour and allow yourself to get a good night's sleep every night if you want to increase your ability to produce.

All of the items in this chapter may seem like common sense, although many people in this day and age neglect themselves, expecting their bodies to be able to achieve excessive amounts of work without the right fuel to keep the body and mind going. That's unreasonable and increases the stress levels.

The habits that are shown in this chapter are important ones. The physiological state of your body depends upon the correct food, sufficient drink and also sufficient sleep. If you deprive yourself of any one of these, you will find that you are quickly going to get tired during the day. The reason that I thought it was important to include the breathing exercises is that if you can use deep breathing, you can help yourself to focus at work and not to panic when you have bitten off more than you can chew.

I know how easy it is to concentrate on the amount of papers on your desk, rather than concentrating on what you are doing. If you can breathe correctly, you stop yourself from hyperventilating and are able to cope with whatever the workplace throws at you. It's a very valuable lesson to learn and one that you need to practice in order to make it a part of your life.

CHAPTER 4 – HABITS TO AVOID IN ORDER TO BECOME MORE PRODUCTIVE

In the opening chapter of the book, or the introduction, I talked about perfectionism and also about fear of failure. The psychological approach that you have toward your life in general and toward your work dictates the outcome. The world isn't perfect. Thus, if you are a perfectionist and expect everything to be done in a certain way, chances are you will be very disappointed because life doesn't work like that.

Perfectionism

When you are a perfectionist, you waste time because you are too busy thinking of the theory of doing a job rather than doing it. You also have unreasonable expectations of people who work with you. Unless you can drop this perfectionism, it gets in the way of productivity. Nothing you do is ever good enough and even when you learn new approaches to work, chances are that you will find fault because that's what perfectionists do. They plan to do things but rarely achieve them. In fact, some don't even start their work because their perfectionism stops them in their tracks. Look at it logically, if you expect perfection, you are going to be very disappointed in life. Sometimes, you can achieve good results but the motivation for doing so should not be perfectionism. It should be flexibility in approach and ability to work with others.

More often than not perfectionists have an unrealistic picture of what needs to be done and this needs to be put into perspective. If you suspect that you are putting off work because of perfectionism, you need to ask yourself the following questions:

- Does perfectionism get in the way of productivity?
- Does perfectionism make you afraid of failure?

The answer to both of these is probably "yes." Thus, you need to accept that life isn't perfect and start to break your work down into manageable proportions without taking your analysis too far. You spend more time analyzing than actually doing. Look at your list of things that you have to do today and see how many of them you have detailed. Now write yourself another list, merely telling you what the job is and how long you think you should give to that task. Make a point of aiming your perfectionism toward accurate ability to judge how long a job takes. This will give you the incentive to prove yourself right and this may just be the kick that you need in order to perform. Stop working out all of the nitty gritty and start doing the job to the best of your ability.

Stop expecting people around you to adhere to your perfectionist standards because you will alienate people and make them less motivated to produce or to achieve. There's nothing worse than working for a perfectionist.

Try giving praise where it is due when people help you to achieve goals.

Being afraid of failure

This aspect of non-achievement is huge. You would be surprised how many people don't achieve because they don't even try. They are afraid of failure but what failure is isn't as bad as everyone makes out. Failure acts as a lesson and when you make a mistake, you need to change your attitude. Stop recoiling into your shell and use the mistake to learn from. It isn't that you:

Failed to do the job properly

It is that you:

Learned how not to do the job, so that you know better for your next attempt

Richard Branson, the multi-millionaire owner of the Virgin group of companies was asked once about whether he had ever failed. He laughed at the question and told the reporter that he fails all the time, but that's how he learns and how he makes new discoveries in the future – by avoiding the same action. You can do the same thing too. Stop letting fear hold you back. It's not healthy.

Once you are able to let go of fear, you stop procrastinating and that's when you see changes happening. You may not know this, but people who are known to procrastinate have two choices:

• Start the work
• Find a reason not to start the work

The second choice will always be the choice of the procrastinator. He will find things to do that seem more important. He will justify not doing what he needs to do by appearing busy doing other things. What you may not realize is what stops people from procrastination. If you start something, you will always think of it as unfinished until you actually get on with it. If you haven't started it, you can drop it from your mind and not even think about it. Thus, you need to get a job started because until you finish it, it will weigh heavily upon you and it's a natural thing for people to want to finish something they have already started. The first push may be the worst one to make, but once a procrastinator makes that move, everything changes.

Other reasons for procrastination should also be addressed. If you don't know how to do something, ask someone. Stop hiding the fact that you don't know because no one will respect you when you have to eventually admit that you didn't know how to do something and people are only too happy to share knowledge. I once worked with someone who hid the fact that she didn't know how to input amounts onto the computer. She was supposed to tally everything that she did and double check but she hadn't learned how to do it and simply shuffled papers across her desk. By the time we found out that she hadn't been doing the job, we were up to our eyes in trouble. She ended up crying and explaining that she was scared of the changes that had been introduced and that she was finding it hard to keep up with it all. Not only did we have handbooks for everyone to refer to, but we openly encouraged people to admit when they needed help. The problem was that no one did ask and she felt that we would consider her stupid if she asked too many questions. Once you do ask, take notes and if there is something you don't understand, ask again. It is far better than you make a good job of learning than simply avoiding asking the questions you need to in order to get the job done.

CHAPTER 5 – GIVING YOURSELF MOTIVATION

People who are productive and who are also successful don't rely upon others to validate them. For instance, I know that, as a writer, I have to produce a set number of words a day. Otherwise, it may as well be a hobby. I set myself goals just as I have suggested that you set yourself goals, but the mistake that some people make is needing validation from others. When they don't get it, they lose motivation.

You need to find that motivation inside you. I do this by being happy to break my own records. For example, if I aim at 5000 words today and write 7000 tomorrow, it motivates me and makes me feel productive. If I do the bare minimum, I tend to see that effort as mediocre and chances are that you will feel the same way as well about relatively small goals. However, what you are not seeing is the big picture. When your list for the day is finished, you free yourself up to do other things. Think how nice it would be to be able to offer someone else in the office a little bit of help. How about being able to go home early and spend more time with the children? You have to see the benefits of your work yourself, rather than depending upon validation from others. This is what gives you incentive and motivation.

When David started working in an insurance job, he got pretty bored with the routine of the work. Sure, telephone calls broke this up a bit, but the majority of his work was routine. He knew that his only route to promotion was to show his efficiency in handling the work and also handling the people who worked under his supervision and devised a competition for himself whereby he would compete with himself to get things done in less time. He also instilled this kind of mentality in his staff and they were able to keep on top of routine work and free up more time to offer to do other work that was more interesting. In fact, his team impressed the manager so much that David was promoted to handling staff training and managed to instill that same sense of importance and efficiency into all of the staff he trained.

There's something very special when you are able to say to yourself:

"I did it!"

I have seen this work time and time again in giving people the incentive to work hard. People who had very little motivation were given the challenge of setting themselves goals and these were marked on a large chart at the front of an office. Each time that a member of staff managed to reach one goal, he/she had to mark it on the chart to show their progress and what happened was that the competitive nature of humans kicked in. Everyone wanted to be a winner. You can actually do this on your own without everyone in the office seeing your personal chart but you get the same buzz from it. By making your own personal goal list and measuring your work in such a way that you fill your day with productivity, you can feel that same sense of satisfaction at the end of the day when all the items are ticked off.

Start it now. Don't delay it. I personally have 16 items to do today and I know that I will do them. In the past, I may have worried about having that many things to do, but I am motivated because I see it as a personal contest with myself. I don't do it to prove to others that I can. I do it to up the ante and see how much I can manage to do and love the challenge that this gives me. Set your own challenges. Make tick lists but don't spend all day concentrating on lists. Then, start to work your way through the lists and see how it feels when every item on the list is completed. There are those who won't appreciate what you do, but that's because they are usually too busy doing their own thing, but learn to do it for your own appreciation, rather than for the approval of others. One thing I have noticed is that people with self-esteem issues go above and beyond to prove to others that they can succeed. When you know you can succeed and you don't have to prove it to anyone else, you become much more valuable as a staff member because that means that you can work on your own and get things done without a fuss. This is much more in line with management and will help to get you promoted much faster than seeking approval all the time, which makes you look needy. Needy people don't usually get management posts.

Speeding yourself up

PRODUCTIVITY

I personally love this exercise and feel that it achieves so much. When you have a pile of work to do, use the alarm system. Set the alarm for 45 minutes and work solidly during that 45 minutes, knowing that at the end of this time, you can stop and have a walk around and enjoy coffee or your favorite beverage without feeling guilty. When your break is over go back to the timer and set another 45 minutes of solid work during which time you give your attention to the work and nothing else. People around you will get used to you using this system and will respect you when they see just how much you are capable of producing. It's almost like magic. What you are doing is focusing yourself on one thing during the work time and then giving yourself a break at the end of it. After you have had three 45 minute sessions and are due another break, make this a longer break, such as an hour for lunch. You will have deserved it and you can enjoy sitting down and actually eating your lunch instead of trying to eat a sandwich while you work through your lunch hour. Why it works is because you are setting yourself challenges and working all through the 45 minutes, you achieve that challenge and actually produce more work than you thought you were able to.

I use this system when I have deadlines. I work out what the deadline is and exactly what I need to achieve within that deadline and then split the tasks down to simple tasks that I can tick off as and when I achieve them. By doing this and using the alarm, I am concentrating my attention on one job at a time and not allowing myself to multi-task. Thus, I am not asking my mind to skip from one thing to another, but putting my full attention on each of the smaller tasks that make up the whole task. If you try this system, you will find that you will up your production by around 50 percent and that's a pretty amazing achievement. One word of caution, use a vibrating alarm rather than one that upsets the whole office!

Using this system, you are able to achieve so much. All of your concentration goes onto the job at hand and you don't let other things get in the way of achieving that job. If I am bothered by someone during what I call my busy time, I simply ask them to come back later when I have finished what I am doing. No boss on Earth will object to that once he sees the amount this ups your productivity.

The tips in this chapter will help you to speed up the process of getting through your work, even on days when you feel that there's more than you can handle. You have two choices. Pile up all the easy work and get that done quickly to half the load, or work on the hardest items first while you have the mental energy to do that and then have the rest of the day easier. It's really up to you and what kind of character you see yourself as being.

Personally, I prefer to get rid of the hard jobs first, knowing that nothing for the rest of the day will give me so much bother!

CONCLUSION

Throughout this book, I have encouraged you to encourage yourself. At the end of the day, that's the only person you can depend upon. You need to know that your lack of productivity is a result of YOUR approach. It isn't the fault of the person who gave you the work. It isn't the fault of someone who didn't do his share. When you are willing to take responsibility for your own lack of organization and are able to apply the above exercises to your working life, you will find you get more out of yourself and feel more motivated to keep working at that particular level of production.

Another thing that is worth remembering is that unhappy people don't work to their full potential. In other countries this is thought of as of paramount importance and employers go out of their way to make the lives of their workers pleasant. In the western world, however, it is left to the employee to find his own incentive. It's a good idea to remember that you only have one life. You need to balance it in such a way that you have all the elements of WORK, REST AND PLAY. That means making sure that you have family time. It means giving your personal life the same level of importance as your home life. There's not much point in earning millions if you never have the time to share them with your son or daughter and don't really have the energy to participate in his or her life. It's not worth being so unproductive that you lose your job either. You need to find the balance.

Although you may think that you are indispensable, the fact is that no one is. When you die, it's forever. You only get this one chance to balance your home life and your work life and if you can find that balance, you give yourself all of the incentive you need to be productive both at home and at work and can use the strategies given in this book for both your home life and your work life. Start to see things in perspective and realize that your lack of productivity spells your lack of sensible approach. Change your approach and up the incentive and begin to enjoy your life to the fullest.

Made in the USA
Columbia, SC
21 February 2019